WAPUSK

White Bear of the North

Jillian,
you are my
angel in my heart
love Mom
May 2018

WAPUSK

White Bear of the North

Photography by Dennis Fast

Main text by Rebecca L. Grambo

Heartland Associates Inc.
Winnipeg, Canada

Printed in Manitoba, Canada

Left: Few people have ever seen a polar bear loafing in a bed of blazing fireweed. Maybe that's because few people have ever camped in the presence of a bear so relaxed in his summer environment. The bear opposite spent all day only about thirty metres from our small compound on a remote island along Hudson Bay's northwest coast. Moments earlier, he'd come in contact with the electric fence that protected us and the jolt made him jump back in a hurry. Twice he circled the fence, looking for an opening, but he never again touched the wire. Then he lay down to watch developments and invited us to come out and play!

National Library of Canada Cataloguing in Publication Data

Fast, Dennis. 1943 –
Wapusk: white bear of the north / photography by Dennis Fast;
written by Rebecca L. Grambo

ISBN 1–896150–32-2

1. Polar bear - - Pictorial works.   2. Polar bear.   I. Grambo, Rebecca, 1963
II. Title.
QL737.C27F37   2003   599.786   C2003-910369-2

Heartland Associates Inc.
PO Box 103, RPO Corydon
Winnipeg, MB R3M 3S7
hrtlandbooks.com

Perfectly adapted to their environment, polar bears often nestle into a snowdrift to sleep, sometimes waking to find themselves wholly or partially buried, as this youngster was.

CREDITS

PHOTOGRAPHY and
MUSINGS BEHIND THE LENS
Dennis Fast

MAIN TEXT
Rebecca L. Grambo

EDITOR
Barbara Huck

DESIGN
Dawn Huck

RESEARCH & PHOTO ASSISTANCE
Frieda Fast

DRAWINGS
Barbara Endres

PREPRESS
Avenue 4, Winnipeg, Canada

PRINTING
Friesens, Altona, Canada

The long, warm days of late June and early July carve magnificent sculptures of sea ice.

TABLE OF CONTENTS

PREFACE

I cannot really recall a time when I was not fascinated by the natural world around me. From early childhood, I dreamed of being a bird photographer. Little did I realize that what would develop from that dream would be a dual passion for birding and the photographic pursuit of nature.

Today I search for beauty everywhere, in the incredible close-up world of window frost, or the majestic grandeur of a polar bear at home in his arctic kingdom. In either case my goal is the same: to capture in a split second the haunting beauty of nature in fragile tension between the present and the inevitability of change.

This book is about passion, beauty and wonder: passion for a natural world often threatened by man, a celebration of the beauty that can be found anywhere by those who have eyes to see, and not least about wonder, the sense that somehow the polar bear and his world were created to send shivers of awe through our being. It is humbling to walk in his steps for a while.

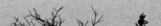

Battered by harsh northern winds and challenged by summers just three months long, the black spruce of the transitional forest, or taiga, left, are hardy survivors. Growing from a root system that may be hundreds of years old, what appear to be clumps of black spruce are often in fact just one tree. And despite its barren appearance, this is a region filled with life, including the beautiful snowy owl, above right, which can be seen year round in northern Alaska, as well as along the western shore of Hudson Bay and the north coast of Ungava.

8

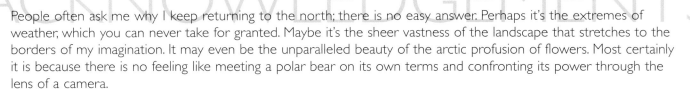

ACKNOWLEDGEMENTS

People often ask me why I keep returning to the north; there is no easy answer. Perhaps it's the extremes of weather, which you can never take for granted. Maybe it's the sheer vastness of the landscape that stretches to the borders of my imagination. It may even be the unparalleled beauty of the arctic profusion of flowers. Most certainly it is because there is no feeling like meeting a polar bear on its own terms and confronting its power through the lens of a camera.

Thank you to everyone who had a part in this book: Peter St. John and Barbara Huck, who encouraged me; Dawn Huck for designing such a beautiful book; Rebecca Grambo for her excellent text; Pat Price for his brown bear photo and his portrait of me, and for shared photo adventures; my wife Frieda, for supporting me over many years, my family, Stuart, Byron and Val, and friends too numerous to mention, without whom I could never have done this.

And especially I need to thank a group of friends who have shared "critique" nights with me for more than twenty years. You have consistently inspired me and pushed me to new levels of artistic insight. Last, but not least, thank you to Mike and Jeanne Reimer for making Churchill my home away from home. I would have no polar bear stories without you.

Dennis Fast
April 2003

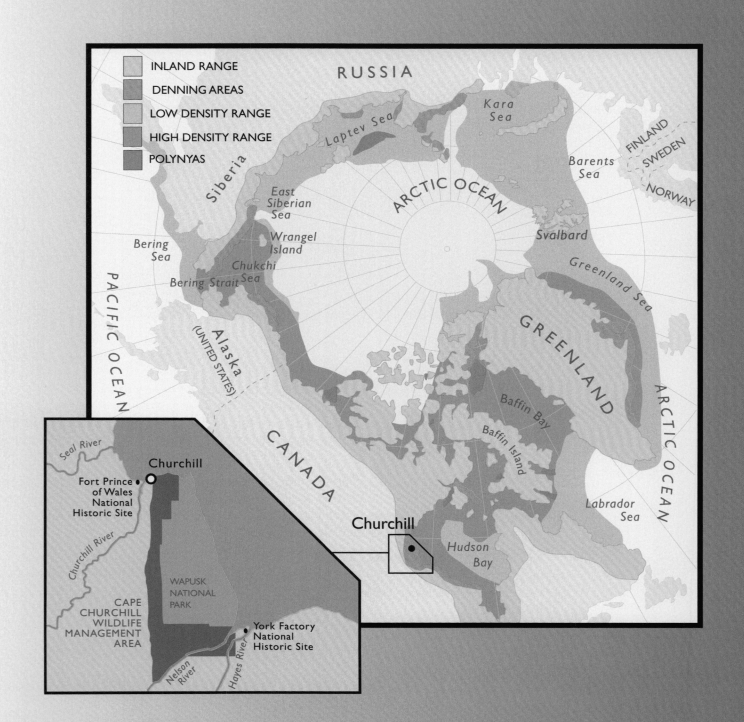

INLAND RANGE
DENNING AREAS
LOW DENSITY RANGE
HIGH DENSITY RANGE
POLYNYAS

RUSSIA

Kara Sea

Laptev Sea

Barents Sea

FINLAND

SWEDEN

NORWAY

Siberia

East Siberian Sea

ARCTIC OCEAN

Svalbard

Bering Sea

Wrangel Island

Chukchi Sea

Greenland Sea

Bering Strait

GREENLAND

PACIFIC OCEAN

Alaska (UNITED STATES)

Baffin Bay

ARCTIC OCEAN

CANADA

Baffin Island

Churchill

Labrador Sea

Hudson Bay

Seal River

Churchill

Fort Prince of Wales National Historic Site

Churchill River

CAPE CHURCHILL WILDLIFE MANAGEMENT AREA

WAPUSK NATIONAL PARK

York Factory National Historic Site

Hayes River

Nelson River

Ancient rock and crystalline water combine to create a serenely beautiful natural sculpture.

An October frost turns the northern taiga into a symphony of blue and silver.

12

DEDICATION

For all those who have fostered my love of nature, for those who share it, and for those who may be inspired to a deeper awareness of the beauty of wild things through my images.

13

The polar bear rules a kingdom dominated for much of the year by snow and ice. Wapusk – "white bear of the north" – as the Cree named him, not only survives but thrives in one of the world's harshest environments. How might such an animal have evolved? Digging into the fossil record for clues to the bear's past allows an exploration of the remarkable adaptations that enable it to succeed in the unforgiving arctic regions of the world.

About sixty per cent of the world's estimated 22,000 polar bears call Canada home. The rest live in Norway, Russia, Greenland and the state of Alaska. However, as a symbol of one of the last, vast wild spaces on the planet, the polar bear has been adopted by people around the world. This global treasure and its arctic habitat are under pressure from an increasing number of sources. Using a combination of my text and Dennis Fast's personal recollections, this book will examine those threats and look at what can be done to counter them. It will also attempt to paint a picture, with words and images, of one of the world's most majestic creatures.

Rebecca L. Grambo
April 2003

Whether loafing among boulders at the edge of the sea, or posing on the ancient granite shoreline of Hudson Bay, polar bears dominate their arctic homes.

It is bitterly cold. A rising wind drives sharp ice crystals before it, blurring the landscape. Suddenly he's there, little more than a swirl of snow, moving on great, furred paws. He pauses, lifts a black nose and questions the wind, then changes course slightly. The curtain of blowing snow thickens and the polar bear vanishes, once more becoming part of the land that has shaped him.

16

Evolution

THE LAND AND WATER TRULY DID make the polar bear what it is today. The late Finnish paleontologist Bjorn Kurtén theorized that polar bears evolved from a population of brown bears that became isolated from other brown bears by a glacial ice sheet sometime between 100,000 and 250,000 years ago. In an environment where there were plenty of seals but no land predator eating them, the bears evolved rapidly to take advantage of the empty niche. Individuals that possessed advantageous characteristics, such as warmer fur, had a better chance of surviving to reproduce. Selection under such extreme conditions produced not just a different brown bear but eventually a new kind of bear altogether.

Several lines of evidence support this theory. Polar bears are a very young species, likely even younger than Kurtén hypothesized. The oldest bones found so far that can be called "polar bear" are about 100,000 years old, while DNA studies indicate that modern polar bears may have appeared only about 20,000 years ago. As little as 10,000 years ago, polar bears still had many brown-bear-like molars, and were not yet meat-eating specialists. Recent DNA studies show that the DNA of brown bears on Admiralty, Baranof and Chichagof Islands of southeastern Alaska is more like that of polar bears than of other brown bears. This not only confirms the close relationship between the two species, but may indicate that polar bears evolved on this side of the Bering Strait.

The evolution of the polar bear from the Alaskan brown bear, above, happened so recently that the two species can still interbreed and produce fertile offspring. This likely never happens in the wild, but has been documented in captivity.

18

Physical features are one issue; behavior patterns are another. Could a land-dwelling, omnivorous brown bear gradually change its lifestyle to become a carnivore so at home in the water that its name, *Ursus maritimus*, means "bear of the sea"? Researchers and native hunters have seen brown bears, which are good swimmers, hunting seals along the coast and on shore-fast ice. In May 1991, biologist Dr. Mitchell Taylor supplied proof that at least one brown bear saw nothing wrong with travelling out on to sea ice to hunt. Five hundred kilometres north of where he would expect to see one, Taylor spotted a large male brown bear on the sea ice. After tranquilizing and weighing the bear, Taylor followed its tracks back to where it had been dining on ringed seals. Carefully scrutinizing the snow around the seal kills for polar bear tracks and finding none, Taylor concluded that the brown bear must have made its own kills and not scavenged carcasses from a polar bear kill.

Polar bears are perfectly at home in the water, perhaps more so than any other land mammal. They have often been sighted more than fifty kilometres from land and on one occasion, a large male was seen more than 200 kilometres from the nearest shore.

20

The brown bear supplied suitable raw material. From it, the cold ice age environment fashioned a superb predator with paddle-like front paws for swimming, thick, white fur for warm camouflage, and teeth for tearing and cutting meat rather than for grinding plants. The new bear's neck was longer, helping to keep its head up when swimming. A layer of blubber, sometimes more than ten centimetres thick, under the skin added buoyancy in the water and provided such effective insulation that a running bear would quickly overheat. Smaller ears also conserved body heat. The bear's feet grew larger, helping to distribute its weight on thin ice, and the soles became heavily furred. The remaining pads of bare skin on the paws were resistant to frostbite and covered in thousands of tiny bumps, providing excellent traction. Short, strong, curved claws could hook and haul a seal from the water. In short, the new bear that evolved was a hunter perfectly designed to spend the arctic winter roaming the sea ice.

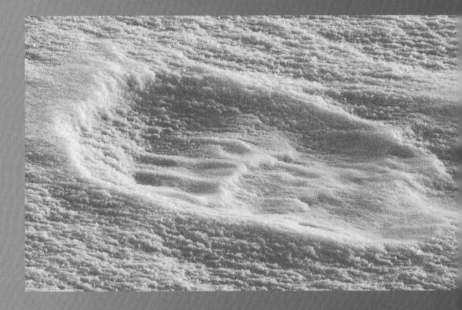

This print of a big male in the snow was so large that a grown man was able to fit both snow boots inside it, with room to spare. In general, polar bear prints are much larger than those of a grizzly or brown bear. Brown bears have longer claws, however, which are adapted for digging; the polar bear's shorter claws allow better traction on the ice.

The Perfect Predator

The bear and the seal were in the water, about sixty metres apart. Each time the seal surfaced, the bear paddled slowly towards it; when the seal submerged the bear floated motionless in the water. Each time the sequence was repeated, the bear drew closer. Finally, the seal surfaced right next to the bear and was instantly killed. Witnessing the entire performance, a small group of spectators could only speculate that the seal had believed the bear to be a harmless piece of floating ice.

AS A PREDATOR THE POLAR BEAR WAS highly respected by early northern hunters. In many areas, bears and humans sought the same prey, using skill and intelligence to secure the food they needed to survive. For both, this generally meant catching a seal. But seals are naturally cautious and to be successful, a hunter must be exceptionally patient and cunning. Many legends relate how early arctic peoples learned their seal-hunting skills by watching polar bears. The bears, of course, have advantages not shared by humans, including an extremely acute sense of smell. They are able to locate a seal beneath its lair of snow and ice up to a kilometre away.

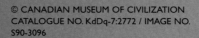

Once their prey is located, polar bears usually use one of two basic hunting strategies. The most common is still-hunting: the bear simply remains motionless beside a seal lair or breathing hole and waits for its prey to come within range. The other technique is stalking, which involves a stealthy approach followed by a sudden leaping plunge to seize the seal. Stalking is commonly used during the whelping season, when seals are giving birth and nursing their pups in snow lairs. If a scent tells a bear a lair is occupied, it will cautiously approach, then plunge through the roof to seize the seal within.

Still-hunting and stalking generally work well, but in some situations, polar bears have been known to develop variations that give us an idea of just what "thinking" predators they are.

The ivory seal, opposite, may have been carved nearly 2,500 years ago. It was discovered on Baffin Island.

23

Ringed seals use the short, sharp claws of their front flippers to create and maintain three or four breathing holes in the ice. If snow covers the hole, the seal often scratches out a small snow cave so that it can haul out on the ice without being seen. A polar bear will wait, often for hours, lying on its stomach near a breathing hole for a seal to appear. It will invariably position itself downwind of the hole because seals, with good reason, are wary and immediately take a quick sniff when they emerge from the water. The bear listens for the sound of the seal's breathing to signal its arrival. If the ice is thin, the bear will then smash through with its paws, often crushing the seal's skull with the same blow. When faced with thick ice, a bear may prepare ahead of time by digging around the breathing hole to weaken the ice. Or it may dig a separate hole some distance away. When it hears the seal, it will slip into the water and grab its prey from beneath.

ged seals, below, and their young, which are born in March and early April, constitute one of the main rces of food for polar bears. Six-week old seal pups gh about thirty kilograms, of which half to three-quar- is fat. Dining on such delicacies, polar bears can put iterally hundreds of kilograms in a matter of months.

Open water requires slightly different techniques. A bear may either dig a hole in the ice near the water's edge, much as it would near a breathing hole in thick ice, or it may simply wait near the shore. Humans have also found that polar bears have an uncanny ability to hide in plain view on the ice, along a rock-strewn shore or even in a field of flowers. In any case, when a seal appears, a bear will slide almost soundlessly into the water to catch it.

Seals sleeping in the water, noses pointed to the sky and bodies rocking gently with the waves, provide easier prey. Moving with great care, polar bears can sometimes swim or stalk close enough to the unwary sleepers to kill them. The same is true when seals sleep on small, flat pans of ice or along the shoreline edge of the ice, though the technique is somewhat different. Swimming toward the seal, the bear will silently submerge while still some distance away. Continuing underwater, the bear will surface next to the ice pan, leap onto it and cut off the seal's escape. Author and naturalist Richard K. Nelson suggests that bearded seals must be quite vulnerable to this type of attack because they are such sound sleepers and often lie on very small pieces of ice, allowing a bear to swim very close before jumping out to surprise them.

Both polar bears and the seals they prey upon need sea ice for hunting, resting and reproducing.

Sometimes polar bears display a level of cunning that surprises even researchers who know them well. For example, seals resting on the ice next to their breathing holes sometimes get a nasty surprise: a polar bear suddenly pops up out of the hole, cutting off any chance of escape. Polar bears also stalk seals by swimming under the ice from breathing hole to breathing hole, taking a breath and checking their bearings at each hole before diving again.

Once it has killed a seal, a polar bear usually eats only the skin and blubber. This is where most of the seal's food value lies: the blubber from a yearling ringed seal may contain nearly seventy per cent of the total calories in the carcass. As a bonus, these calories are in a concentrated form, which allows the polar bear to eat less food and still gain weight. One researcher calculated that the average active adult polar bear would require either two kilos of seal blubber or five kilos of seal muscle to fill its daily caloric needs. An exception to the "blubber is best" rule can be seen in cubs still with their mother. Eating the muscle she leaves behind, they get the protein they need to grow.

According to Canada's Ian Stirling, who has spent thousands of hours observing polar bears, the bears surface, breathe, lift their heads to look, and go under again without creating so much as a single ripple in the water.

Along with polar bears, bearded seals face an uncertain future if current climate trends continue. Walruses, too, rely on sea ice as resting platforms between dives to the sea bottom, where they feed on clams, crabs and worms.

Choosing blubber over red meat may have another, less obvious benefit. The metabolic wastes produced during blubber digestion are water and carbon dioxide, which the bear's body can use or get rid of by exhaling. If the bear eats food higher in protein, such as seal muscle, its body produces urea, which must be excreted in urine. The bear must replace the water lost in this process, possibly by eating snow. People who get stranded in winter are advised to avoid eating snow because it takes so much body energy to melt the snow into water. The same is true for bears. A polar bear who eats seal muscle rather than blubber gets less calories per bite, and, if forced to eat snow to replenish body fluids, must burn some of those calories, which reduces the meal's value even more.

Polar bears eat prey other than seals, including walrus, muskox, carrion and, every once in a while, an arctic fox that has become too bold in snatching up the bear's leftovers. Polar bears will also kill beluga whales, particularly when belugas become trapped in a polynya, a small area of open water surrounded by ice. Off the west coast of Alaska, polar bears killed forty belugas under these conditions. No one witnessed the killings but observers later counted about thirty bears feeding on whale carcasses spread on the ice around the open water. Occasionally, a polar bear may do a little foraging on the tundra and eat berries, grasses, kelp and bird's eggs, as well as the odd lemming or vole, but it is seal hunting that keeps a polar bear fed.

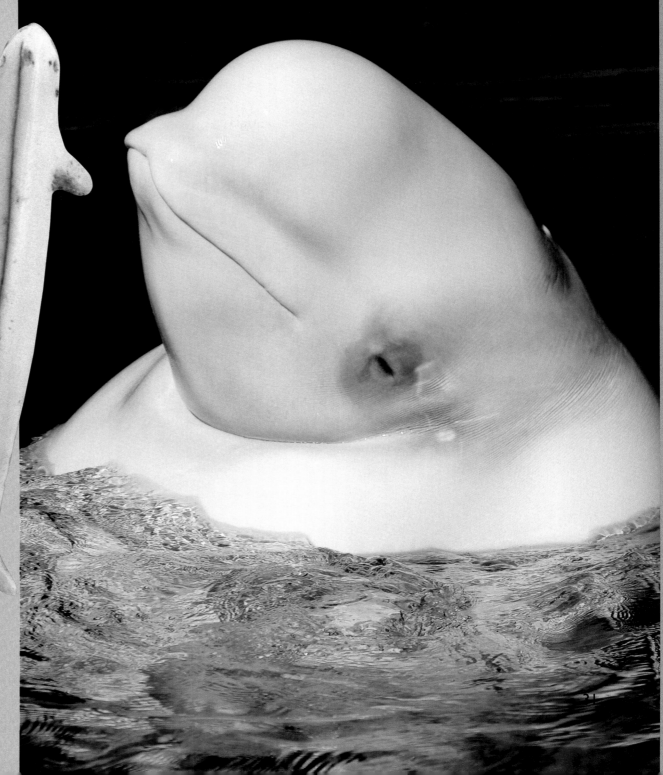

In addition to polar bears, the Arctic is also home to belugas — from the Russian for "white whale" — the most vocal of the whale species.

That white whales have fascinated people for millennia is evident in the discovery of this beautiful ivory beluga created about 1,500 years ago by a Dorset carver.

In July, after the ice breaks up, they gather by the thousands in the river estuaries of Hudson Bay to feed. An estimated 3,000 return each year to the mouth of the Churchill River and even more can be found at the mouth of the Seal River to the north. In both estuaries, tour boats equipped with hydrophones cruise among pods of snowy white adults and their gray-colored young (new-born infants are brown), eaves-dropping on their underwater songs.

The curiosity is mutual, apparently; belugas are occasionally spotted spy-hopping, right, to get a good look at the tourboats.

Not being much of a diver, I took these two photographs of whales in captivity.

Basalt outcroppings along Hudson Bay are softened in summer by an ever-changing palette of wildflowers. White Arctic avens, left, are among the earliest to flower, and provide a counterpoint to masses of sweet-smelling purple vetch. But botanists and photographers are warned not to wander here without protection, for polar bears, particularly the big males, spend the summer close to the bay. Females and their young tend to summer farther inland.

During their enforced period of summer fasting, polar bears do not eat much. They may catch an occasional unwary seal, or stumble on a beluga whale carcass and feast for a while. More likely, though, summer means a sparse diet of birds' eggs and young, some berries or even kelp, along with a refreshing drink of water.

North of 70° north latitude, polar bears often follow the receding ice edge in the summer. Along the shores of Hudson Bay, however, large numbers of bears come ashore, and move slowly southward. The pattern is reversed in October, when they move north to meet the newly-formed ice on the bay.

Birder's Paradise

I have watched polar bears on several occasions saunter through a colony of Arctic terns, perhaps with a feast of eggs in mind. Each time, the bear was harassed by the unrelenting birds, which flew without fear at its head. And each time it was so preoccupied with warding off angry terns that it forgot about finding something to eat.

Birders travel thousands of kilometres to see the graceful terns, as well as rare Ross's gulls and some 200 other species that breed each spring along the shores of Hudson Bay. The tidal flats and salt marshes just west of Cape Churchill are famous breeding grounds for snow geese, which are now so numerous that they are chewing up the tundra. And along the coast everything from Arctic loons and tundra swans to peregrine falcons can be found.

Also found are, clockwise from top left: the northern hawk-owl, a small, tame owl, here at the limit of its range; a Bonaparte's gull, the only gull that regularly nests in trees; Canada goose goslings in huge numbers, and a Hudsonian godwit in breeding plumage. **37**

Battling Bears

IN THIS REGION, POLAR BEARS MATE in April or early May. Males battle fiercely for the right to breed with females, and the numerous scars on some old males are silent testimony to these annual contests. The winning male mates with the female repeatedly and then leaves to resume hunting. Once copulation is complete, so is the male polar bear's contribution to cub-rearing. The female goes on her way, bent on continuing a period of conspicuous consumption that may double or triple her weight, building the fat reserves that will sustain both her and her cubs.

© CANADIAN MUSEUM OF CIVILIZATION
CATALOGUE NO. QiLd-1:2299 / IMAGE NO. S90-2906

Early people who shared the white bear's domain had enormous respect for it, and often carved its image in ivory, bone or wood. This ivory bear, found on Bathurst Island, was created by a Dorset carver about 1,000 years ago.

A good-natured wrestling match between a pair of two-year-old siblings, right and left, is broken up by a larger, more aggressive male.

Autumn contests between males are much less serious, more like play. The same two big, sleepy males now drowsing side by side may have spent the last few hours wrestling and shoving each other around. A male that wants a wrestling partner will approach another male about the same size. He may swing his head from side to side. The bears nose and gape at one another, jaws spread wide and often pressed mouth to mouth. Eventually the play starts in earnest and the two bears grapple with one another, each trying to gain an advantage. They may stand on their hind legs or roll about on the ground. Much of the action seems to be in rather lazy slow-motion, but some blows are delivered with a lightning fast paw. The game continues until either decides to quit, usually for a nap.

Polar bears go through a "play-fighting" ritual every fall as they wait for the ice to form on Hudson Bay. Sometimes, as is the case with these young bears just learning to fend for themselves, the fighting becomes quite intense. Even so, serious injury is rare and only a few blood stains reveal that some of the nips were intended to hurt.

41

Sometimes one partner wants to keep playing. One spectator watched two large males half-heartedly wrestling one morning. One of them quit and lay down for a nap, his head pillowed on his paws. The other bear sat and studied the situation for a bit. Then, without warning, he stretched across and chomped his companion in the posterior. The resulting melee threw clouds of snow and bear breath into the air for a good fifteen minutes before both bears stopped to rest.

Mother Knows Best

TUCKED SNUGGLY INSIDE A SNOWDRIFT-covered den on the lee side of a hill, a female polar bear drowses as her two tiny cubs nurse. Each weighing a little more than a pound of butter and only about thirty centimetres long, the cubs will grow quickly on their mother's creamy, fishy-tasting milk. It is over thirty per cent fat, eight times as rich as human milk. By the time she leaves the den in March to hunt for seals, the cubs will be able to follow her out onto the ice. For now, they rest against her warm fur, unaware of the remarkable biological events that brought them into being.

Almost 200 females den every year in Wapusk National Park, west and south of Churchill on Hudson Bay, giving it North America's largest concentration of polar bear dens.

The highest density denning area in the world is on Siberia's Wrangle Island, where scientists have located 500 maternity dens.

Maternal pride and childlike devotion are written all over the faces of this mother and cub. And who could tell that this idyllic scene was photographed at −49°C?

45

This mother and her cub-of-the-year (COY) have just emerged from a snow den, after leaving their birthing den a few days before. The cubs — a second cub is behind the mother — are only about ten weeks old and are just beginning to face the harsh reality of life in a frozen lanscape. Barely heavier than a house cat, statistics tell us that only one will survive the first year of life. The other could grow to be the largest carnivore on Earth.

Within the female's body, the fertilized eggs have not behaved as they would have in most mammals. Instead of immediately implanting in the uterine wall to begin growing, the eggs are "placed on hold". In a process known as "delayed implantation", the eggs do not resume development until late August or early September. At that time, the female's body condition determines how many cubs she will bear or whether she will bear any at all. If she is carrying enough fat, the usual two eggs will implant themselves. In a very exceptional year, she might have three cubs. If she has been unable to gain sufficient weight, only one egg or even none at all will develop. Any unused eggs are reabsorbed by her body. Delayed implantation increases the probability that only females with good odds of raising healthy cubs will have them.

Few mammals rival polar bears as mothers. Alone with as many as three cubs, they must nurture, educate, defend and provide for their young in one of the most hostile environments on earth.

Each autumn, while the rest of the 1,000 polar bears in the southern Hudson Bay population travel to the coast and out onto the ice, pregnant females seek out safe den sites. In the high Arctic, polar bears usually den in snowdrifts along hills and mountains near the sea. Some may choose to use snowbanks out on the sea ice itself. Each female typically digs a tunnel into a large snowdrift, and excavates two connected rooms at the end. With her claws she smoothes the interior of the farther, larger room and thins the roof to admit a little light and air. This main chamber sits higher than the entrance tunnel, trapping any warmth generated by the bears inside. In Wapusk National Park, south of Churchill along Hudson Bay, polar bears dig earth dens into the permafrost on slopes that will later receive a good insulating snowdrift. Once the drift forms, the female underneath digs out into it, adding on to her den.

Making the den comfortable and well-ventilated is important to the mother-to-be. She will stay here, without eating, from the time she enters the den in the fall until she leaves with her cubs in the early spring.

Pregnant females living near Hudson Bay have been known to survive in a den like this one without food for up to eight months — the longest known fast of any mammal.

Gentle and solicitous, and incredibly tolerant of their offsprings' playful nips, polar bear mothers are also surprisingly non-threatening to humans who keep a respectful distance. They simply do not want a confrontation, for a fight might mean exposing their young to danger.

51

Although she will not truly hibernate, her body temperature will drop by a few degrees and she will spend most of the time sleeping. Once her cubs are born, they will also sleep through most of their first winter, conserving energy and making the most of the calories they draw from their mother's limited body resources.

A polar bear mother is hungry when she leaves the den in the spring, but she travels slowly at first so that her cubs can keep up. The cubs have no protective fat layer to keep them warm so they often snuggle beneath their mother's fur to sleep. If the bear family must cross water or go through deep snow, the mother carries the cubs on her back. Once out on the sea ice, she will periodically dig a hollow in the snow where she can nurse the cubs and they can all sleep out of the wind.

A mother polar bear is intensely protective of her offspring, always on the lookout for males that could pose a threat. Her aggressive defence probably developed as a result of the wide open environment in which polar bears live. Black bear mothers send their cubs scurrying up a tree at the first sign of danger, but female polar bears do not have that option. As the white bears were evolving to survive in the treeless arctic, aggressive females that fiercely defended their cubs raised more off-spring and passed along their beneficial belligerence.

Polar bear cubs are completely dependent on their mother and stay with her for at least eighteen months. The cubs nurse throughout this time and share the seals their mother kills. They watch her hunting behavior closely, learning the skills they will need. Weaning takes place soonest in southern habitats where it is easier for the bears to make a living. Farther north, where survival is more difficult, the cubs may stay with their mother for up to three years. When the time comes for the female to wean the cubs, she abandons them or actively drives them away. This abrupt change in her behavior baffles the cubs, which often linger nearby for some time before finally wandering away.

Some young bears do not learn their lessons well enough before leaving their mother's side; starvation is the leading cause of death for subadult polar bears, those that are on their own but are less than five or six years old.

Once a polar bear reaches adulthood, having mastered the knack of arctic life, its chances of continued survival are excellent. Wild male polar bears often live to be about twenty-five, while females may live slightly longer. One captive female celebrated more than forty birthdays, but a wild polar bear more than thirty years old is rare.

Polar bears remain on the sea ice as spring and summer progress, sometimes travelling hundreds of kilometres to stay with their prey as the floes retreat northward. The amount of food they take in during this period is crucial to their survival. When the ice breaks up, bears are left stranded on shore. Denied a hunting platform and with a poor chance of catching any seals, they have little choice but to begin a long, drowsy fast that will last until the ice forms once again.

Prelude to a Requiem

Mid-October brings snow, bears and visitors to the edge of Hudson Bay. The visitors are there to watch for bears and the bears are there to wait for the ice and the snow. For those of us with cameras, a fresh snowfall creates refracted diamonds that glitter on the tundra and decorate the willows; for the bears, who have fasted since June, it's a harbinger of much-needed months of feasting.

Though they are regular visitors at this time of year, the bears do not appear on command. Lodges in the area must provide alternative entertainments for guests, some of whom have travelled halfway around the globe. One such diversion is a hike into the nearby transitional forest for a bannock roast over an open fire. As we're about to leave for the hike, someone spots a polar bear across the lake behind some willows. Binoculars appear and soon it's clear there are two bears, barely visible, moving behind the ridged edge of the lake. Another ten minutes and a mother and cub move into full view and she heads down the esker toward the lodge. Suddenly, a second cub appears.

Moments later the bears have gathered beneath the picture window of the lodge as shutters click and video cameras whirr. The excitement is quickly tempered, as we realize that the mother is emaciated and in very poor shape. We wonder whether her rather large satellite collar has kept her from eating properly last winter. It certainly seems that it might have impeded her attempts to catch seals at breathing holes. Whatever the reason, she looks gaunt. But the cubs seem to be fine; she has looked after them well. After a few minutes, she turns and leads them back across the lake, where she settles down in the willows to rest.

The next morning, they are still there and in the early morning light, we photograph them framed by fresh hoarfrost. Only the arrival of a helicopter bearing new guests sends them hurrying out of sight.

They return in the afternoon, but this time their visit is complicated by the appearance of a rather aggressive young male, perhaps five years old. Late at night, the family again makes its rounds, then retires just thirty metres from the window.

Early the next morning, after a stretch and a stroll, the mother settles in a hollow, with the cubs cuddled next to her. Their faces are framed by willows covered with hoarfrost, and backlit by the rising sun. It's a beautiful scene; sadly, I am now sure it is a prelude to a requiem.

At noon, while we are waiting for lunch, the cubs suddenly saunter through camp. I can see the mother slowly shuffling though some willows, as the cubs encircle the entire lodge, wandering well out onto the lake. Suddenly, they realize they are farther from their mother than ever before and panic sets in. They run in opposite directions, rearing up on their hind legs and peering about. Even when they spot her, they seem not to recognize her. The larger cub circles downwind and, catching her scent, dashes to join her. But the smaller cub is in a panic. He bounds through camp, paying no attention to me as he races by. All the while he is barking and mewing like a lost puppy, running in huge circles as he smells the air and the ground for clues to her whereabouts. When he finally spots her, he seems not to be able to believe his eyes. He circles once more and, finally picking up her strong scent, rushes to join his family. Peace slowly returns, but now I can easily imagine the even more traumatic experience that awaits the cubs.

The following day, one of our party spots the trio as he walks through the willows. Mother does not move, even though he is just metres away before he sees her. Another day passes, with no sign of her. Now, we cautiously approach. She lies, her head stretched out, the snow beginning to accumulate on her neck and back. The cubs are still leaning on her for support, but there is no warmth there, no leadership to determine their next move. They cling to her, grief and anxiety apparent on their faces.

I photograph the family for the last time and contemplate what the future holds for the cubs. Without their mother, their chances for survival are considerably reduced. At worst, they face a slow lingering death by starvation, or a quick losing battle with a large male, who will target them to ease his own hunger. Their chances of finding enough food to sustain both of them, and avoiding the perils of life on the bay are slim. Statistics say that at least one of them will not make it through the winter. Mother Nature is cruel, only to be kind.

It is rare to witness the death of a polar bear. Only two, that I know of, have been found. The rest disappear, perhaps dying out on the ice to be swallowed by the sea, perhaps quickly salvaged by other creatures in need. We finish the morning with a walk to the coast and then I sit down to try to put the morning's experience in perspective.

Animals die every day. Right now, some polar bear mother is struggling to meet the needs of as many as three cubs. But it's easy to rationalize, easy to intellectualize, not so easy to put this moment out of mind. The death of a magnificent animal is always deeply moving, perhaps because it is such a powerful reminder of our own mortality. May I die so eloquently.

The following day, a Natural Resources helicopter arrives to deal with the dead mother bear. As the confused and terrified cubs run across the lake, the helicopter lifts her body and heads north, carrying it up the coast. The rest of the day, the cubs repeatedly come through camp and wander the lake and surrounding tundra. This brings a series of standoffs with another bear that is also in the area.

At dinner, one of the cubs suddenly appears at the dining room window, peering longingly at the bountiful food on the table. We all feel both compassion and guilt, but we must drive it away, for we can not encourage its presence near the lodge. Already the cubs seem bolder and more aggressive without their mother to caution them, and the smaller one takes a series of runs at one of the staff members, as he pumps water from the lake. Though the cubs are not yet two, they are as large as an average-sized black bear, and could certainly be dangerous.

Epilogue: The cubs remained around the lodge until November, dodging large males and trying to interact with other mothers and cubs as they moved through the area. Natural Resources officers attempted to relocate the cubs out on the newly-formed ice on Hudson Bay, but they quickly found their way back to the lodge.

They were still there when the lodge closed for the season. Whether they survive or not depends on the lessons they learned from their mother. If she had begun to teach them to hunt during their first winter on the ice, they have at least a fighting chance of survival. If not, they are surely doomed. Their mother, it was determined from information gathered from her radio collar, was twenty-four, a ripe old age for a wild bear. She did her duty to the very end.

60

Mother, may you rest in the peace
Of all wild things,
And may your children live
To bear your courage.

61

Past & Present Challenges

1100 A.D., Arctic Canada.
The tracks in the snow were crisp and fresh. Despite his warm fur clothing, the Inuk hunter shivered. He knew he was close to his quarry and understood the encounter he was seeking would mean death for one or the other. He knew, too, the grave spiritual danger he risked, for the polar bear's spirit was very powerful. He looked down at the stone-tipped spear he carried and thought of the small ivory charm that lay against his skin. He would count on the power of both to keep him alive today.

This spirit bear, carved about 2,000 years ago, may have been created to protect a Dorset hunter as he faced the daunting Arctic.

In the past 250 years, over-hunting has posed the main threat to polar bear populations. In the last half-century, however, new problems have become evident. Increased human contact, pollution, and global warming are all adding their weight to the polar bear's burden.

Direct human impact on the global polar bear population began when early northern hunters first started regarding the white bear as prey rather than predator. Hunting polar bears is therefore not something new, but methods, motivations and numbers of bears killed have changed dramatically over time.

Only female bears can be radio collared, because the diameter of an adult male's neck is bigger than that of his head, allowing a collar to easily slide off. Males are ear tagged instead.

Native hunters had an overwhelming respect for the polar bear, not only because it was a physically dangerous quarry, but because it held strong spiritual qualities. From their first encounters, man has admired the great bear's hunting prowess. Turning that admiration into food production was a risky business. Aua, an Iglulik Inuk, explained to Danish ethnologist Knud Rasmussen:

"The greatest peril of life lies in the fact that human food consists entirely of souls. All the creatures that we have to kill and eat, all those that we have to strike down and destroy to make clothes for ourselves, have souls like we have, souls that do not perish with the body, and which therefore must be propitiated, lest they should revenge themselves on us . . ."

It was once believed that polar bears wandered the arctic regions at will. The truth is more remarkable. Though they travel large distances — even a mother with young cubs can cover thirty kilometres a day — bears belong to distinct populations and stay in the same region, sometimes despite the rapid movement of ice. In the Beaufort Sea, off northern Alaska, the Beaufort Gyre moves the sea ice at speeds of up to several kilometres a day. With no landmarks or obvious navigational references, the bears somehow manage to stay in the same place, moving at precisely the same speed, though in the opposite direction, to the flow of the ice.

It was essential that a hunter treat the bear in a respectful manner before and after he killed it so that the bear's spirit might speak well of him. Otherwise, the hunter might never again have luck finding game. He could also be made ill or even be killed by a bear's angry spirit.

The polar bear's *innua*, or soul, was believed to be more powerful than that of any other animal. To give the innua time to prepare itself for what was to come, a hunter would customarily make several passes over the bear's carcass with his skinning knife before making the initial cut. For five days after a female bear was killed (or four days for a male), the bear's innua lingered dangerously on the tip of the hunter's spear. To prevent the bear's spirit from becoming "crooked" or evil, people did no work inside the house. They hung the bear's skin outside in a place of honor, surrounded by offerings of tools. If the bear was female, women's tools such as sewing equipment and skin scrapers were set out. For a male bear, knives, harpoons and other men's tools were carefully placed. The people also laid offerings to the bear's spirit on the skin. The dances and gift-giving ceremonies that followed a successful polar bear hunt sometimes continued for more than a week.

BARBARA ENDRES

Heading out onto the ice to hunt, this polar bear mother leads her two small cubs past three innunguait or stone sentinals on Baffin Island.

65

The traditional harvest of polar bears originally posed no threat to polar bear populations. Inuit hunters went after polar bears with spears: five- or six-foot-long shafts tipped with jade, copper or perhaps a flake of meteoric iron. They killed relatively few bears and there were no other significant threats to the population. Bear skins were not acquired for sale or trade but only to satisfy local needs for food and clothing. The situation changed substantially with the arrival in the Arctic of non-native commercial and "sport" hunters.

Starting in the 1700s and continuing through the mid-1900s, hunters using the latest in modern technology came looking for adventure, profit and polar bear skins. They killed large numbers of bears and as new equipment appeared – snow machines, airplanes, rifles with telescopic sights, even set gun traps – in many areas the take became too great for the polar bear population to sustain.

An animal that roams large areas of inhospitable, largely inaccessible wilderness belonging to several countries is difficult to census accurately. But even without solid numbers it became clear by the mid-twentieth century that something had to be done to stop the unregulated slaughter. In 1965, scientists gathered at the first International Scientific Meeting on the Polar Bear. Soon afterward, the Polar Bear Specialist Group (PBSG) of the International Union for Conservation of Nature and Natural Resources (IUCN) was formed to oversee international polar bear research and management.

67

In 1973, the five polar bear nations signed the Agreement on the Conservation of Polar Bears, which came into effect three years later. This unique document set out an ecosystem-based approach to conserving polar bears and their habitat. Denning areas, migratory routes and feeding areas were recognized as critical regions. The agreement committed the five countries to sharing research data and managing shared bear populations, using sound conservation principles based on the best currently available information. Finally, all five countries agreed to ban the hunting of bears from aircraft and large motorized boats.

Reaffirmed unanimously in 1981, the agreement has been a crucial and effective tool in the understanding and protection of polar bears and their arctic habitat. A new conservation agreement, signed by the Russian Federation and the United States in 2000, seeks to improve management of their shared polar bears by placing some provisions of the International Polar Bear

Agreement into law. The agreement also includes provisions for the active involvement of Native people in management decisions. Russia has passed legislation ratifying the treaty, while in the United States it is now being considered by the Senate Foreign Relations Committee.

October on Hudson Bay

Two annual migrations occur in October on Hudson Bay's west coast. The southern polar bear population on the bay begins to move north en route to Cape Churchill and people from around the world arrive daily in Churchill to watch them. More than 15,000 visitors arrive each year. Despite the concentration of both species in a relatively small area, hazardous encounters are few. Strict regulations, public education, bear patrols, and a relocation program for problem bears minimize the dangers for both bears and humans in Churchill itself. Out on the tundra, a limited number of specially designed vehicles carry bear watchers on day trips out to where the bears are. Real enthusiasts can make arrangements to stay in a special lodge on the tundra near Churchill, or even in one all the way out at the Cape. Visitors out on the tundra are not allowed on the ground; it's amazing how a white bear can remain unseen in brown willows until you're nearly on top of it. No feeding or baiting of the bears is permitted at any time, a regulation strictly enforced to protect the bears.

The polar bears seem unconcerned about all the human interest, usually ignoring the buggies full of picture-taking people and settling down for naps. Some bears are curious enough to initiate contact, walking right up to the vehicles for a good look, a sniff and sometimes a taste of the tires. Mothers bring their cubs for a look but usually keep them from getting too close. Eventually, the number of bears seen near Churchill begins to drop as more and more of them work their way out toward the Cape.

Looking more like a passenger train, a multi-car tundra buggy, opposite below, heads for Cape Churchill; when polar bears are sighted, its windows, left, and rear viewing area, below, suddenly bristle with camera lenses.

AS RESIDENTS OF CHURCHILL, MANITOBA can attest, people love polar bears. The annual influx of tourist dollars accounts for sixty per cent of the town's economy — a powerful reason for protecting the annual influx of bears. But this was not always the case.

Churchill was founded in 1771 as a fur trade post and in the early years polar bears were killed as food for sled dogs. When an air base was built east of Churchill during World War II, servicemen wanted to take home the highly valued pelts. "Nuisance" bears were usually shot, until 1982 when the Polar Bear Jail was established to hold bears until they could be relocated away from the town. In 1983, bear hunting became illegal. Today, residents are quite tolerant of the bears and the occasional property damage they inflict. While not perfect, Churchill's experience with bringing bears and people together can serve as a guide for other areas where ecotourism is a burgeoning business.

"Problem" bears are sedated and driven to "jail", a facility capable of handling sixteen individuals and four family groups. There, they are held until freezeup, when they are released back onto the ice.

76

If the holding pens become overcrowded, some of the inmates are taken by helicopter at least thirty kilometres along the coast.

A growing number of tourists are travelling to remote areas, such as Svalbard, on the Arctic Circle north of Norway, to see the bears, and it is important that governments, scientists and tour operators work together to avoid potential human-bear conflicts. It is also important that human impact on the extremely fragile arctic environment be kept to a minimum. No one – tourist, tour operator or researcher – wants to see polar bears and their habitat loved to death.

Thanks to a unique combination of wind and water circulation patterns, the sea ice forms at Cape Churchill earlier than anywhere else in the region. Each autumn as the weather turns cooler, bears – mature males, subadults, mothers with cubs – appear in the Churchill area as if by magic. The adult males, who have spent the summer along the coast, move out to the Cape and do their waiting there. Mothers with cubs hang back and stay out of the males' way.

Polar bears spend more than eighty per cent of this time on land sleeping and resting. Since they are not eating, or at least not much, it is important that they conserve energy. They enter a physical state scientists call "walking hibernation", which allows them to survive a summer without food – something that would starve a black bear. Near Churchill, mothers and cubs can often be spotted curled up together so tightly that it is difficult to tell where one bear ends and another begins. Large males can also be seen resting together, an unusual thing at any other time of year for this normally solitary animal.

While waiting for the bay to freeze, the bears wisely conserve energy by lounging and sleeping. Polar bear expert Ian Stirling has shown that under normal circumstances, the bears will come ashore in June with as much stored fat as they need to survive until late October. The exceptions seem to be mothers with nursing cubs and young bears that may not yet be skilled hunters.

79

Ready for the icy blasts in its winter plumage, the willow ptarmigan winters in deep thickets or near the shoreline of the bay, where it can find its principal food — willow leaves. In summer, the bird dons a plumage of rusty brown.

Out on the Cape, the bears patrol the shoreline restlessly. When the tide retreats, some venture out onto the young ice, checking its extent and thickness. As the tide comes in, the bears return to shore. Occasionally observers at Cape Churchill may see a bear waiting, motionless, out on the ice for a seal to surface, or bears and ravens clustered around a patch of red snow that signifies a kill.

And then, suddenly, the bears are gone, except for a few stragglers. Once the ice will bear their weight, the hungry hunters waste no time. The coming winter will bring many challenges for all polar bears, especially for mothers with cubs to feed, but thousands of years of evolution have prepared them to face most challenges and prevail.

Some threats facing polar bears, however, may prove too difficult for them to overcome without help – help that must come from the origin of those threats: humans.

Walking Hibernation

I've spent a lot of time at a polar bear lodge near Dymond Lake, which is strategically located west of Churchill on an esker overlooking Hudson Bay. Troublesome bears that are flown from "jail" in Churchill are often deposited only ten or twelve kilometres along the coast from the lodge. As a result, they frequently wander by, giving us ample opportunity to observe their behavior.

Almost every year, a bear or two "adopts" the lodge as a good location from which to wait for the bay to freeze. Sometimes, a bear even manages to catch an unwary seal as it naps on the rocks below the esker at low tide. Perhaps the bears like the lodge, which incidentally was named for an early scientist and not the jewel, because of all the interesting human activity and the gourmet smells emanating from the kitchen. Whatever the reason, many bears seem to love to visit and

linger for a while, particularly if no big male has claimed the spot.

During the summer and late fall, polar bears are in a state of what is sometimes called "walking hibernation". They don't hibernate in any real sense, but loaf a lot and sleep away many hours of each day. The places they choose for these naps are as varied as the bears themselves.

They do love comfort, however, and that comfort includes a pillow, if it's available. You and I might not think of a rock as a great support for our heads while we snooze, but I have seen polar bears do just that. When an appropriate rock is not available, a chunk of ice will do. So will a willow bush or an elevated tussock on the uneven tundra.

If there's enough snow around, bears will scoop out a real bed, fit for a monarch of the wild, and it will almost always include a snow pillow, piled to a comfortable height. Then they really snuggle in for a while.

Bears also choose some odd places to bed down. It's not unusual for them to spend considerable time sleeping in full view of the lodge, and occasionally even resting for a time on the landing of the front steps during the night. We wouldn't let them get that close during the day, but the night is the bear's domain. We do try to keep a close watch for bears at night, to prevent damage to the buildings or outright break-ins. That sometimes means frequent trips around the compound to make sure all is well.

One year we had a pair of huskies that alerted us to every bear with short repeated woofs, and they were always right. One night, however, we couldn't find any bears, in spite of incessant barking by the dogs and many tours of the yard on our part. In the morning, we discovered that the bear had been sleeping inside a caribou skin tipi that we'd walked past many times during the night. The dogs knew he was there, but then, our sense of smell is not as good as theirs.

In October 2002, we had an ailing mother bring her twin cubs to the lodge. The very first night, they slept on the edge of the esker, only twenty metres from the picture window. We were able to take photographs of the family using flashlights and I even took some pictures utilizing only the light of the full moon.

For the most part, bears sleeping anywhere within sight of the buildings (or out of sight, for that matter) choose to stay downwind. That way they can absorb all the smells as well as hear what is going on while they doze. Even the resident cubs of 2002 exhibited this instinctive behavior; when their mother died and left them orphaned, they would almost always sleep on a ridge at the edge of the lake just south of the lodge. There they were downwind of the prevailing north wind and the lodge, and had an unobstructed view over the lake to the southeast. In this way, the major areas of potential trouble were all covered.

From Hunter to Hunted

BETWEEN 500 AND 700 POLAR BEARS, or approximately two or three per cent of the total population, are killed each year. Canada, Alaska, Greenland and Russia allow hunting, following guidelines in their international agreement. Most of the bears are harvested by indigenous people, who include the hunt as part of their cultural tradition. Remote Native communities may also gain important income from the sale of bear hides.

Canada is the only country that allows hunting of polar bears by non-Native sport hunters. Hunting licences are issued under quota systems to Native groups (except in Quebec and Ontario, which do not set quotas). The groups can then decide how many licences, if any, to sell to non-Native hunters. The income generated by a sport hunt is much larger than the value of the single polar bear hide. If a sport hunt is unsuccessful, the licence cannot be reissued, and the polar bear kill limit for the community is effectively reduced.

Svalbard, Norway has no aboriginal community that can claim traditional harvesting rights and, since 1973, polar bears in Norway are a protected species. They may be killed only in self-defense, to protect property, or as a mercy-kill. Polar bears in the western Russian Arctic receive the same protection, but poaching is a potential problem.

Traditional harvest by local people appears to be sustainable in most cases, but there are concerns about kill numbers in some areas. In Greenland, for example, regulations are intended to allow only Inuit subsistence hunters to kill polar bears, but there is no quota system and the population status is not well known. Close monitoring of bear populations and cooperation between government and Native groups will ensure the preservation of both bears and Native traditions.

Poisonous Pollutants

IN THE MID-1980s, UNIVERSITY OF LAVAL SCIENTIST Eric Dewailly was studying levels of chemicals called PCBs in the breast milk of women from southern Quebec. He wanted an uncontaminated population to use as an experimental control. Believing that communities in the pristine Arctic, far removed from the sources of these pollutants would be ideal, he began testing inhabitants. What he found astounded the scientific community. Levels of PCBs in Inuit mothers were as much as five times higher than those in women from the south. The most likely source was the "country food", such as seal and whale, that the Inuit women had eaten.

PCBs belong to a class of chemicals known as persistent organic pollutants (POPs). You have probably seen the names of some POPs in the news over the years: dioxins, furans, lindane, dieldrin and, perhaps most infamous of all, DDT. They are herbicides, pesticides, industrial chemicals and industrial by-products. The Inuit breast milk study was not the first indication that POPs were present in the Arctic – by the mid-1970s DDT and other pesticides had been found in seals, polar bears, beluga whales and fish — but the new study showed that POPs were reaching humans. Suddenly, the race was on to learn more about where these chemicals were coming from and what effects they might have on humans and wildlife.

Local sources, such as military installations, mining operations, and local pesticide use, account for some of the POPs found in the Arctic. The majority, however, come from distant sources, such as industrial and agricultural chemicals used in southern industrialized countries. Prevailing air circulation patterns carry POPs to the Arctic and concentrate them there.

The "persistent" in POP comes from the fact that they hang about in the environment, resisting breakdown, for a very long time. Many of them have an affinity for fat and are stored in the fatty tissues of the body or passed along in fat-rich fluid, such as breast milk. In the arctic regions, where many animals depend on fat for insulation, POPs are exceptionally dangerous for

top-level predators because they bioaccumulate as they are carried up the food chain. What seem like infinitesimally small amounts detected in sea water can become very significant in animals near the top of the chain. Theo Colborn, in *Our Stolen Future*, gives an example: a small shrimplike animal called a copepod eats a bit of plant that contains PCBs.

The copepod is consumed, along with many others, by a hungry cod. The PCBs from this meal are added to the others stored in the cod's fat, raising the concentration of PCBs in the fat to [...]ty-eight million times that of the surrounding [...]. A ringed seal gobbles up the cod, and hun-[...] f other fish throughout its life, storing [...]ccumulated load of PCBs and other chemicals. The seal's blubber may have eight times the PCB level found in the cod – and 384 million times that of the water. The blubber-loving polar bear gets that concentrated dose of PCBs each time it eats a contaminated seal. The bear's PCB level may reach three billion times that of the water. What does all this mean for the bear?

Research indicates that high levels of PCBs may adversely affect regulation of thyroid hormone levels, reproductive success, and immunological response. If the polar bear is a nursing mother, her high-fat milk delivers a concentrated dose of chemicals each time her cubs suckle. And PCBs are just one of the POPs found in increasing levels in arctic regions all around the globe.

Beautiful and delicate, the purple paintbrush grows on sandy beaches and gravel ridges along the shores of Hudson Bay and James Bay south of the Arctic Circle.

One of the most disturbing qualities of POPs is that many seem to mimic reproductive hormones, causing impaired breeding behavior and various reproductive disorders in several species. No definite link has yet been demonstrated for polar bears, but scientists have made a disturbing discovery on Svalbard, where PCB levels in bears are high. Some female polar bears, four out of 269, displayed the malformation known as pseudohermaphroditism – the partial formation of both male and female genitals in the same animal. Studies are ongoing on Svalbard and in Canada to examine the relationship between POPs and polar bear reproduction. The PBSG of the IUCN is also conducting a circumpolar study to determine current contaminant levels in polar bears, which can then be compared to the results from a similar study done ten years ago. This will determine trends involving previously found pollutants and identify new ones.

Besides POPs, the polar bear faces other types of pollution. A growing host of new, man-made chemicals, including flame retardants, are also concentrated in the Arctic. The potential effects of many are simply unknown.

Heavy metals, such as mercury and cadmium, have been found at high levels in both seals and polar bears. Mercury acts as a neurotoxin and can have a negative impact on the brain development of bear cubs. It can also disrupt sperm production in male bears. There is no evidence that heavy metals are threatening the polar bear population as a whole. However, current knowledge about the sources and effects of heavy metals in arctic regions is limited.

Fat soluble pollutants pose a special risk for gestating female bears. Because they fast during gestation, they are burning fat and therefore increasing the concentration of these pollutants. Many of these chemicals have been shown to disrupt the hormones that affect the development of embryos, as well as the later survival of the cubs.

Along with Russia, Denmark and Canada, Norway has taken steps to protect maternity denning areas in their national territories. Kõngsõya, on the Norwegian territory of Svalbard, is one such area.

Radiation is not a problem one normally associates with the Arctic but many reactors – including those on ships and submarines and large amounts of nuclear waste are located there, especially in northwestern Russia. The overall radioactivity level of the arctic marine ecosystem has increased since the Nuclear Age began. Any large release of radiation would affect the marine food web and its top predators, including polar bears.

Is there any good news about pollution and polar bears? Yes, levels of some chemicals that have been banned in many countries seem to be starting to decline. More chemicals may disappear in the next decade.

On May 22, 2001, delegates from 127 countries voted approval of the Stockholm Convention on Persistent Organic Pollutants. Each government must now vote on ratification of the Convention and fifty must ratify it for it to become international law. It's a process that's expected to take at

least five years. The Stockholm Convention is aimed at the elimination of some of the world's most hazardous chemicals, including PCBs, DDT, dioxins and furans, through scheduled phase-outs or outright bans.

The arctic regions are rich in economically desirable resources, including petroleum and minerals. Exploration for and development of these areas has the potential to harm polar bears and their habitat.

Seismic blasting, helicopter, boat and vehicle traffic, as well as construction of facilities are all part of exploration for oil and minerals. In sensitive areas, such as polar bear denning sites in Alaska's Arctic National Wildlife Refuge, increased human activity could disturb or displace bears.

Oil and related products pose big threats to bears, particularly in the event of a marine spill. Oil reduces the insulating value of a polar bear's fur, causing it to require more calories to keep warm. Scientists have concluded that having oiled fur for even a short time can kill a bear, mostly by poisoning it as it tries to clean itself.

This ancient ivory swimming bear almost perfectly echoes the real thing.

Bears can also be poisoned by eating oil-coated seals or sea-birds. A major spill near one of the prime denning sites, such as Hopen Island in the Barents Sea, could affect a large portion of the world's polar bears.

There is no evidence to date that existing resource developments have had any large-scale effects on the polar bear population. This is prob- ably due to the fact that obvious conflicts have been avoided and only limited development has happened in crucial polar bear habitats. However, as resource development reaches ever farther into the polar bear's sensitive environment, it is impor- tant that precautions be taken to minimize impact, and that plans be in place to deal with possible environmental emergencies.

On Thin Ice

GLOBAL WARMING IS A PHRASE THAT seems to be on everyone's lips these days – especially if you talk with polar bear researchers. The IUCN's Polar Bear Specialist Group believes that climate change poses one of the biggest threats to polar bears. An increase in temperature can dramatically alter the bear's habitat and its crucial hunting platform, reducing the thickness of sea ice and shrinking the area it covers in more northerly regions. Farther south, it also shortens the length of time between ice formation and break up. The data are still coming in but it seems clear that changes are already happening in some arctic regions.

Though this juvenile northern shrike looks too young to be deadly, he will soon be preying on small mammals and birds that inhabit the taiga.

Over the last 100 years, temperatures in the north have risen by approximately five degrees Celsius. The area covered each year by sea ice has decreased by about six per cent during the past twenty years. Scientists using models to study and predict climate change agree that sea ice will likely continue to decrease over the next fifty years. During the last twenty years, the thickness of the pack ice has decreased, too, from ten to forty per cent depending on how the data is interpreted.

One of the first polar bear populations to be affected by climate change is also one of the most studied, photographed and economically important. The sea ice season is shorter in Hudson Bay than anywhere else in the Arctic Ocean, creating a precarious state of affairs for polar bears. They need every hunting day and every seal they can get to survive the four to eight months they must fast while on land.

In western Hudson Bay, ice break-up now happens about two weeks earlier than it did twenty years ago. Scientists have shown that every lost week of hunting sends the bears ashore weighing about ten kilos less and in poorer condition. For females, this missing fat may be crucial to their cub-rearing success. Studies in the area between 1980 and 1992 showed that an average of fifty-three per cent of cubs survived from spring until the sea ice formed again in the fall. Only thirty-eight per cent of mothers did not lose any of their offspring. Survival appeared to be linked to the cubs' body mass; heavier cubs did better. And studies have shown that heavy cubs usually come from fatter females.

Commonly seen on the tundra, this snow bunting can be identified by its large white wing patches.

The polar bears of Hudson Bay are already living at the margins of polar bear habitat and at the limits of their bodies' ability to adapt. A continued reduction in sea ice could severely impair their ability to reproduce successfully. Ian Stirling and other polar bear experts believe that if Hudson Bay ever stays free of ice for most of the year, polar bears will probably disappear from its shores.

Polar bears evolved to meet the demands of climate. Can they now adapt to a warming world? Some bears seem to be trying. Studies by Ian Stirling and Sara Iverson showed that bears around Hudson Bay have started eating harbor seals and bearded seals, in addition to ringed seals. Bears need ice to get to the ringed seal, and when ice is not available, they must find other sources of food. This is probably only a short-term solution, however.

Global temperatures have risen before during the polar bear's existence. From about 850 to 1250 A.D., northern North America, Greenland, Europe and Russia enjoyed a period known as the "Medieval Warm Period" or the "Little Climatic Optimum". Temperatures believed to range from about 1°C to 3.5°C warmer than today reduced Arctic sea ice and allowed the Vikings to settle what really was a "green" land. Little is known about the effects of the Medieval Warm Period on polar bears, although some evidence suggests that they moved north of their usual range. That would be expected if they were following the retreating sea ice.

Just as less time out on the ice can be hard on bears, an extended hunting season can have positive results. In 1991, Mount Pinatubo erupted in The Philippines, spewing dust and sulphuric acid into the upper atmosphere. The following summer, the ice in Hudson Bay lingered for almost an extra month; the bears came ashore fatter, had more cubs and more of the litters survived.

103

An inukshuk stands against the sky, like the polar bear both a beacon and a symbol of the challenging north.

Around 1200 A.D. the amount of sea ice began to increase around Iceland and Greenland. Polar bear skins, traditionally used for carpeting Icelandic churches, became plentiful there. The Viking colonies on Greenland declined and finally died out around 1500 A.D. The warm period was over and Europe entered the period known as the "Little Ice Age".

The difference between the Medieval Warm Period and the climate change that is happening now lies in the rapid rate at which change is happening and its projected long-term duration. Evolution is a remarkable process but it requires time for species to adapt to changes. And some obstacles, such as a complete loss of habitat, are too great for evolution to overcome. Should continued global warming shrink the arctic ice pack beyond a critical point, which some climate experts feel could happen as early as 2060, polar bears could very well become extirpated in the more southerly regions, and greatly reduced in numbers overall.

Polar bears are a keystone species for the entire arctic marine ecosystem. Balanced precariously at the top of the food web, they act as an indicator of the health of our northern regions. Around the globe they serve as a symbol of this remarkable wilderness and, now, of the dangers that threaten it.

Arctic nations, those most likely to feel the greatest effects of climate change, should be pushing strongly for measures to slow or stop global warming. Three of those nations – Canada, Russia and the United States – are home to most of the world's polar bears. They are also large producers of greenhouse gases. Without a prompt concerted effort by these three countries to meet or beat the standards for reducing emissions set forth in the Kyoto climate treaty, polar bears and their arctic home may suffer irreparable damage. Time is running out.

A white speck in the deep blue waters of Wager Bay resolves itself into a polar bear. The great animal paddles leisurely, its back paws trailing behind, heading towards the nearest large ice pan. Reaching the edge of the ice, the bear effortlessly hauls itself from the water. With a tremendous water-spraying shake, it dries its creamy fur. Pausing, head lifted, the bear looks out across the ice and water of its world. Long ago, they formed a path to what the polar bear has become. Today, they seem to stretch into the future, where an unknown destiny awaits.

Let us hope it is filled with ice and seals for the polar bears, and with polar bears for us.

Polar bears are arguably the world's largest land carnivore. Adult males usually weigh between 400 and 600 kilograms (880 to 1320 pounds), but they can reach 800 kilos. They average between 2.4 and 2.6 metres (7.9 to 8.5 feet) in length. Females are smaller, weighing an average of about 200 to 250 kilos, though pregnant females fattened up for the long denning period are known to weigh as much 400 kilos.

Bibliography

Colborn, Theo, Dianne Dumanoski and John Peterson Myers, 1996, *Our Stolen Future*; New York, NY: Dutton, Penguin Group

Cox, Daniel J., and Rebecca L. Grambo, 2000, *Bear: A Celebration of Power and Beauty*; San Francisco, CA: Sierra Club Books

Ewer, R. G. [1973], 1998, *The Carnivores*; Ithaca, NY: Cornell University Press

Gitay, Habiba, Avelino Suarez, Robert T. Watson, and David Jon Dokken, eds. 2002, *Climate Change and Biodiversity*; Geneva, Switzerland: Intergovernmental Panel on Climate Change (Secretariat)

Kurten, Bjorn, 1968, *Pleistocene Mammals of Europe*; Chicago, IL: Aldine Publishing

Lamb, H. H, 1985, *Climatic History and the Future*; Princeton, NJ: Princeton University Press

Lynch, Wayne, 1993, *Bears*; Vancouver, BC: Douglas & McIntyre

Malcolm, J. R., C. Liu, L. B. Miller, T. Allnutt and L. Hansen, 2002, *Habitats at risk: global warming and species loss in globally significant terrestrial ecosystems*; Gland, Switzerland: World Wildlife Fund for Nature

McGhee, Robert, 1996, *Ancient People of the Arctic*; Vancouver, BC: University of British Columbia Press.

Nelson, Richard K, 1969, *Hunters of the Northern Ice*; Chicago: University of Chicago Press

Norris, Stefan, Lynn Rosentrater and Pul Martin Eid, 2002, *Polar Bears at Risk*; Gland, Switzerland: World Wildlife Fund for Nature

Stirling, I., and D. Guravich, 1988, *Polar Bears*; Ann Arbor, MI: University of Michigan Press

Photographing in the North

I have visited Churchill and other sites on Hudson Bay for many years now, in all seasons of the year. Photographing the sub-arctic has become somewhat of a passion, filled with many rewarding experiences. The North, however, does not give up its beauty without a struggle.

I have made several visits to the Churchill area in March to photograph the northern lights and the bears: the experience is definitely not for the faint of heart. I have made photographs when the temperature was hovering around -40 degrees (Celsius and Fahrenheit are identical at that temperature and feel exactly the same — bitterly cold) while the wind was howling. The wind chill reached -60°C to -70°C for extended periods, making it almost impossible to keep fingers from freezing. Touching a cold camera in those conditions feels surprisingly like touching fire.

Under such circumstances, I would not call the experience "fun"; but when a mother polar bear and her two very young cubs move into your viewfinder after a four- or five-hour wait, it is certainly exhilarating. Equally thrilling is the breath-taking dance of the northern lights (the *aurora borealis*), while you try to fend off the mind-numbing cold.

The wind chill is not only hard on fingers, but on equipment as well. Film becomes brittle and may snap at any time. Interestingly, I have found that my new digital cameras and micro-drives seem to perform better in extreme cold than their film counterparts. Everything slows down, but everything works. The key, of course, is ensuring an adequate, warm battery supply. And it goes without saying that you have to be warm enough to operate the controls of the camera.